ABC of Yellowstone Oddities

An Alphabet of the Obscure,
Endangered, and Underappreciated
in Our First National Park

Written and Illustrated by
Anastasia Kierst

Published by Eternal Summers Press LLC
eternalsummerspress.com
catchingsummer.etsy.com

Copyright © 2022 Anastasia Kierst for text and illustrations
Edited by Anne Victory of Victory Editing
First Edition Published 2022

FOR ROGER,

Welcome to the world! Auntie loves you!
May Yellowstone's wonders capture your
heart and imagination as they have mine.

MANY THANKS

To Yellowstone Forever for giving this book wings by
hosting me as a Yellowstone artist-in-residence.

To Chelsea DeWeese for the generous gift of her time
and expert eye in reviewing my manuscript. Any
inaccuracies that may remain are due to my own error.

"When we try to pick out anything
by itself, we find it hitched to
everything else in the Universe."

- John Muir

A is for Apollinaris Spring.

Cool, clear, and carbonated, I've greeted thirsty park visitors from the very beginning. Early European tourists named me after Germany's famous Apollinaris sparkling spring water. These days it's best to drink only treated water.

SEE SAFETY NOTE.

B is for Bell.

Walking into Old Faithful Inn's big red doors for the first time, you'll probably gaze around, amazed by the place. When you recover, look for me on the back of the front door. One-of-a-kind and handcrafted by a blacksmith, I'm the most unique doorbell you'll ever meet. I'm one more detail that makes the Inn so special.

C is for Corvids.

Brains or bandits, clowns or con artists, we are some of the cleverest creatures in the park! You might not always remember where you left your socks five minutes ago, but we Canada Jays remember the location of thousands of hidden snacks for months. Our raven neighbors will open your pack and help themselves to your lunch.

D is for Death Gulch.

I'm a geothermal death trap. Toxic vapors from my vents can quickly kill on still days. Carcasses attract more victims, creatures of all sizes and kinds, turning my little valley into a barren boneyard.

SEE SAFETY NOTE.

E is for Extremophile.
From ephydrid flies that thrive atop scalding water to microbes and algae that can tolerate an acid bath, we're Yellowstone's most bizarre cast of characters! We're to thank for the rainbow of colors in the park's hot springs and medical breakthroughs. A bacterial enzyme from a Yellowstone hot spring even helped doctors in the fight against the COVID-19 pandemic.

F is for Fairy Slippers.
We are tiny jewels of the forest floor.
Despite our inviting colors, we do not
provide nectar to visiting pollinators.
Instead, we rely on fungi to flourish.
Status: VULNERABLE

G is for Grand Canyon's Hot Springs.

Overlooked in more ways than one, we're the thermal features you can only visit with binoculars! Here, the ancient thermal basin baked the rock, softening it enough for the river to cut the incredible canyon you see before you.

VIEWPOINT: Artist Point, Brink of the Lower Falls, and bottom of Uncle Tom's Trail

H is for Hermit Thrush.

My enchanting melodies seem to float and echo, a haunting song you'll remember forever. You may search, drawn deeper and deeper into the forest, yet never find this mystery serenader.

Status: LOW CONCERN (population stable)

I is for Ice Box Canyon.
My chilly heart is deep and dark,
hidden in a canyon so steep you'll
never see the bottom. I'm hiding away an
icy wonderland that lasts into the summer.

SEE SAFETY NOTE.

J is for Jumping Mouse.

Leaping lizards! I can disappear in a single jump! When danger threatens, I drum my tail to sound the alarm. Then bing! I spring to safety and freeze, so no one sees me. During the short alpine summer, I eat myself silly, packing on fat to survive winter hibernation.

Status: LEAST CONCERN (population stable)

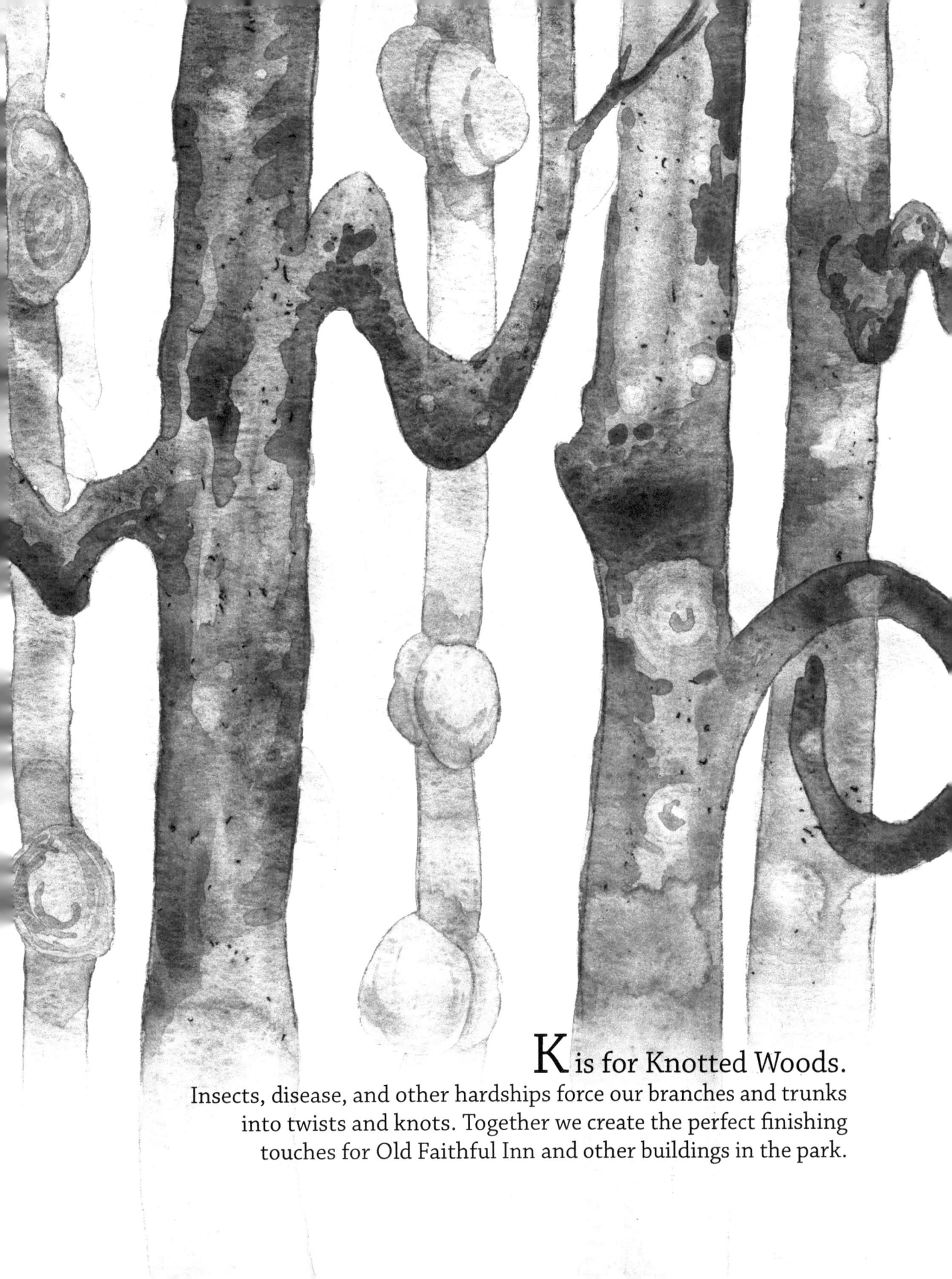

K is for Knotted Woods.

Insects, disease, and other hardships force our branches and trunks into twists and knots. Together we create the perfect finishing touches for Old Faithful Inn and other buildings in the park.

L is for Limber Pine.

I am the pine king—my realm, the windswept and rocky heights where no others can survive. From here, I've watched fires rage and forests smolder. "Better bend than break" is my motto, which has served me well as I've survived for a thousand years. When I'm gone, my bones will be here thousands more.

Status: LEAST CONCERN (population decreasing)

M is for Mycelia.

You might know us better as mushrooms, but those are just the fruits of our labor. We're down here in the dark—tiny fungal threads under your feet, under that log. We are, quite literally, the web of life, linking the living and the dead, breaking down, building up, and unlocking the energy in the world around us. Mysterious and misunderstood, we're more essential than most will ever realize.

N is for Nest.

Precariously perched in this particular place, I've been the first home to many an osprey chick over the years. Millions gaze into the abyss, but only the sharpest eyes notice this nest atop a dizzying drop.

O is for Old Man's Whiskers.
Quietly nodding, we wait for the bumblebees. After they arrive,
we toss our heads back in a puff of "prairie smoke."
Status: LEAST CONCERN

P is for Pronghorn.

I'm a sturdy little speed demon exceeding fifty mph. Why so fast? My ancestors had to outrun the prehistoric American Cheetah! Some call me an antelope, but I'm actually more closely related to giraffes and okapis.

Status: LEAST CONCERN (population stable)

Q is for Queen's Laundry.

Please don't use me as a laundromat or a bath. Those days are gone! What remains is my log cabin bathhouse, the first-ever national park facility built for the public.

SEE SAFETY NOTE.

R is for Riverside Geyser.
I might be one of the most predictable geysers in the world, but that doesn't make me boring! The Firehole River tumbles past, showered by my arching jet. Sometimes I like to throw in a rainbow for good measure.

LOCATION: Continental Divide Trail, halfway between Old Faithful and Biscuit Basin

S is for Spasmodic Geyser.

Variety is definitely the spice of my life! My twenty or so vents spit, spray, spout, or spurt, creating a display of wild activity. Watch my sparkling pools rise and fall as if the earth itself is breathing. Every inch of me is decorated in the most amazing textures.

LOCATION: Geyser Hill in Upper Geyser Basin

T is for Trumpeter Swan.
Yellowstone's warm waters are our winter paradise. More of us used to stick around for the summer, and now biologists would like to know why we're leaving. As the most massive North American waterfowl with a wingspan of up to eight feet, we're hard to miss!

Status: LEAST CONCERN (population increasing)

U is for Undine Falls.

I'm named for wise water spirits. You might catch a far-off glimpse of me from Mammoth, but those who venture closer are rewarded with a close-up view of my multiple plunges. Some have told tales of a hidden passage behind my watery veil.

VIEWPOINT: Grand Loop Road between Lava Creek Picnic Area and Mammoth

V is for Vixen Geyser.

You might say I have a playful personality. Unpredictable is my middle name! I love nothing more than erupting the moment a visitor turns away. Oh, and I'm surrounded by a lovely collection of rocks I've created over the years.

LOCATION: Back Basin in Norris Geyser Basin

W is for Wolverine.

I'm one of the rarest animals in Yellowstone. There may be as few as seven of us in the entire park. I wander far and wide, hundreds of miles per year, in my search for carrion or whatever else I can lay claws on. But I still manage to visit my kits, and sometimes we even head out for an extended family vacation!

Status: LEAST CONCERN (population decreasing)

X is for *Xanthoria elegans*.
But my friends call me the elegant sunburst lichen. When
you see me on a rock, it's a clue that birds or rodents like to
sit there. My claim to fame is the "Tough-as-Nails" award
for surviving 18 months of solar radiation and extreme
temperatures outside the International Space Station.

Status: UNLISTED

Y is for Yellowstone Sand Verbena.

I'm kind of like Goldilocks. Yellowstone is "just right!" I refuse to grow anywhere but this one little acre or so on the shore of Yellowstone Lake. Maybe I just like the warm, thermal sand. Whatever the reason, I'm incredibly rare, so please, watch your step!

Status: UNLISTED

Z is for SS Zillah.

I'm the ghost ship of Yellowstone Lake. I was once a proud little steamer, carrying passengers and even towing a barge full of elk and bison to the zoo on Dot Island. No one is sure what's become of me. Perhaps I'm shipwrecked in the depths of history.

A Note From The Author

Yellowstone! The name alone sends my mind running back through so many happy memories, moments of awe, and adventures! I was a Yellowstone kid. For two precious weeks each summer, we called the park home, sleeping in our tent at Madison Junction. We spent our days roaming the boardwalks, hurrying to catch an eruption, or simply listening to what I imagined was the very breath of the earth at Spasmodic Geyser. We soaked at Boiling River, fished, searched for mushrooms, and paddled our canoe between Lewis Lake and Shoshone Lake. I cried my eyes out watching the park burn in 1988 and celebrated wildflowers and seedlings in the burn scars the following summer.

I was about ten when I met an Artist-in-Residence at the Madison Museum. She explained her process and invited us to try a few watercolor techniques. I never dreamed I would be an Artist-in-Residence almost thirty years later! In 2019, Yellowstone Forever hosted me in their beautiful Yellowstone Art & Photography Center, with Old Faithful erupting just outside the front door. Once again, I had two precious weeks to roam the park, exploring, pondering, taking reference photos, and soaking up inspiration. This book is the fruit of my stay.

You might wonder why I chose to focus on the weird and obscure with so many awe-inspiring, traffic jam-inducing creatures and views in the park. In between the bison, bears, world-famous geysers, and giant waterfalls, I see a place packed with detail and biodiversity that is incredible in its own right! We protect what we love. I hope that in sharing my view of the park with you, you too may love and protect it.

Cheers,
Anastasia Kierst

Safety Notes:

Yellowstone is an amazing place to visit.
However, it can also be a very dangerous place. During your visit, you should be aware of some of the unique risks you might come across.

In thermal areas, you must stay on trails and boardwalks at all times. Many people have been seriously burned or killed when traveling off-trail in thermal areas. Because of this, travel into backcountry thermal areas is now illegal in the park.

For more information on staying safe while exploring Yellowstone, visit nps.gov/yell

Below are some safety considerations for a few places mentioned in this book:

Apollinaris Spring:
Although it looks refreshing, the EPA found contaminants in this spring, and the National Park Service advises modern-day visitors to avoid drinking the water.

Death Gulch:
Being a backcountry thermal area, travel into Death Gulch is dangerous and illegal. However, this little valley can be viewed safely from the trail on a ten-mile, round-trip hike beginning at Lamar River Valley Trailhead.

Ice Box Canyon:
The canyon's walls are incredibly steep, and it runs alongside a very busy road. With these factors in mind, there is no real way to view this canyon. But, it is fun to lower the windows and see if you can feel a temperature change as you drive by! Location: NE Entrance Road between Thunderer Trailhead and Soda Butte Picnic Area

Queen's Laundry:
This area can be viewed safely from the Sentinel Meadows & Queen's Laundry Trail. But remember that hiking off-trail in the hydrothermal areas is dangerous and not permitted. The author would also like to warn visitors to be prepared for mud and biting insects that sometimes make visiting the area a bit of a miserable experience!

Take the Yellowstone Pledge:

"I pledge to protect Yellowstone National Park. I will act responsibly and safely, set a good example for others, and share my love of the park and all the things that make it special."

To Learn More And Make A Difference, Visit:

Yellowstone Forever	yellowstone.org
National Park Service	nps.gov
Junior Ranger	nps.gov/kids/become-a-junior-ranger.htm
NPS Educator Resources	nps.gov/yell/learn/education
Volunteers-In-Parks	volunteer.gov
Artist-in-Residence	nps.gov/subjects/arts/air.htm
Youth Conservation Corps	nps.gov/subjects/youthprograms/ycc.htm
Thermal Biology Institute at Montana State University	https://tbi.montana.edu/-educationmaterials

Conservation Status

Many of the plants and animals in this book are labeled with a conservation status. Here's what those labels mean:

EXTINCT	All of these plants/animals are gone.
EXTINCT IN THE WILD	The only plants/animals left are in zoos or sanctuaries.
CRITICALLY ENDANGERED	These plants/animals will probably become extinct.
ENDANGERED	These plants/animals might become extinct.
VULNERABLE	These plants/animals will probably become endangered.
NEAR THREATENED	These plants/animals might become endangered.
LEAST CONCERN	These plants/animals are common or are not very threatened.
NOT EVALUATED	Scientists have not yet researched this plants/animal's conservation status.

HISTORY & ENGLISH LANGUAGE ARTS EDUCATOR RESOURCE PAGE

COMPARE & CONTRAST

People have been visiting Yellowstone for a very long time. Reread the pages for A, B, Q, and Z. Then do a bit of research about Yellowstone history.
(the park's history page is a great place to look: nps.gov/yell/learn/historyculture/index.htm). Think about your visit to the park. Then write about the similarities and differences between visitors long ago and today. Use the graphic organizer to get started. Questions may have more than one answer.

	Modern Visitors 1915 - today	Early Park Visitors 1872 - 1915	Pre 1872 Tribes, Trappers & Explorers
How did they travel inside Yellowstone?			
Where did they spend the night?			
Why did they visit Yellowstone?			
What kinds of things did they do in Yellowstone?			

SCIENCE and ART EDUCATOR RESOURCE PAGE
Web of Life - Exploring Connections
DIRECTIONS:

1. Introduce the concept of a food chain. Possible examples: algae grow from light and nutrients, insects eat algae, fish eat insects, ospreys eat fish. Also, explain producers (organisms like plants that make their own food from things like sunlight, chemicals, and water) and consumers (organisms that must eat food to survive.)

2. Give each student an enlarged copy of the web of life pieces to cut out.

3. Students are given an organism to write in the center of the hexagonal piece. (Offer choices occurring within the same ecosystem.)

4. Working together and researching when necessary, students attempt to organize their hexagonal pieces and arrows to show not just a simple food chain but a web of life. Arrows are used to show the direction of consumption between organisms.

Note: Expect a bit of chaos. This is a complicated activity. The goal is for students to understand that ecosystems are made up of many connected organisms with complex relationships.

QUESTIONS FOR DISCUSSION:

1. Are there situations where a double-ended arrow would be more appropriate? Where?

2. In an ecosystem, does the web of life have a beginning and ending?

3. What do you think happens if one organism disappears from the web of life? What if many organisms begin to disappear?

4. Do you think people can affect the web of life?

5. Famous naturalist John Muir once said, "When one tugs at a single thing in nature, he finds it attached to the rest of the world." What do you think he meant?

Extension: This activity can be extended as an art project. Discuss abstract and representational art. Choose one style to depict your web of life. Plan and create a mural inspired by the "Web of Life" activity.

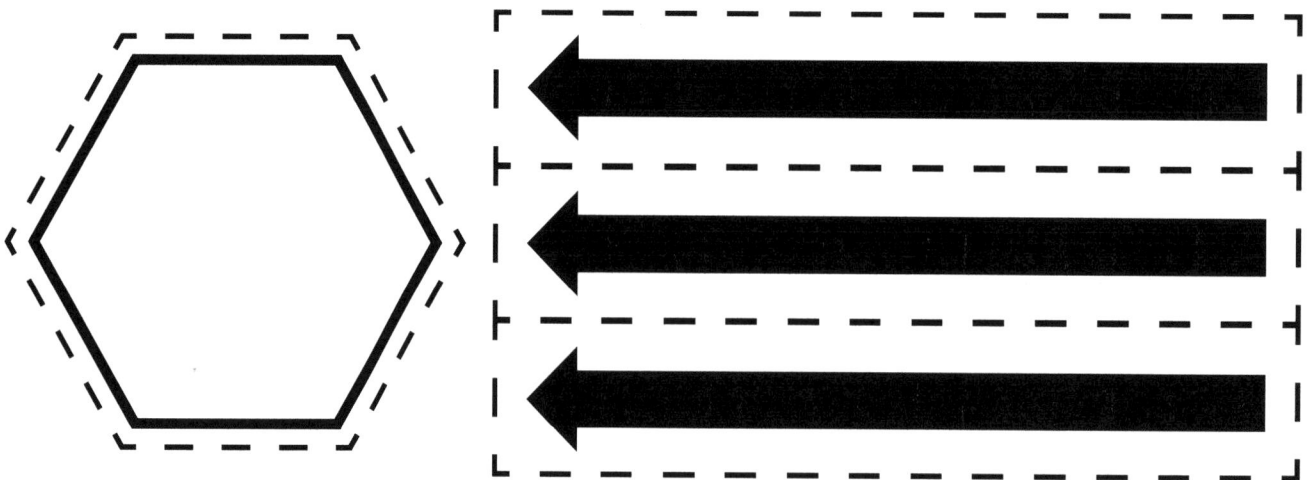

SCIENCE EDUCATOR RESOURCE PAGE

Research and Teaching Others - THREATENED AND ENDANGERED SPECIES
This makes a great follow-up activity to the "Web of Life" activity.

DIRECTIONS: 1. Students, alone or in groups, research a threatened or endangered species in their own geographic area. You might list a few to get students started, or leave that research up to them. Students can complete the graphic organizer below or the one for invasive species on the next page. If you have access to computers, students might use the conservation websites listed in this book to help their research. 2. Students create posters to inform others about their findings and how the community can help.

THREATENED AND ENDANGERED SPECIES		
Choose a local threatened or endangered animal.	List 3 things this animal needs to survive:	1) Habitat — Where does this animal live?
		2) Food Source—What does it eat?
		3) What else? Does this animal have any other special requirements?

What are the biggest threats to this animal?

What can people do to help?

Is there anything people need to stop doing in order to help this animal?

How will other animals, people, or plants be affected if this animal goes extinct?

Why should people care what happens to this animal?

What graphs or pictures can you show on your poster to help make your point?

Research and Teaching Others—INVASIVE SPECIES

DIRECTIONS: 1. Briefly discuss lake trout, how they are hurting the native species in Yellowstone, how they spread, and what people are doing to help. 2. Students, alone or in groups, research an invasive species in their own geographic area. You might list a few to get students started or leave that research up to them. Students can complete the graphic organizer below or the one for threatened species on the previous page. 3. Students create posters to inform others about their findings and how the community can help.

INVASIVE SPECIES

Choose a local invasive species. It could be an animal or a plant.

List 3 way this species hurts native species:

1)

2)

3)

How did this species enter your environment?

How does it hurt people or property?

How fast is this invader spreading?

What can people do, or stop doing, to limit the spread of this species?

Why should people care that this invasive species is spreading?

www.ingramcontent.com/pod-product-compliance
Lightning Source LLC
Chambersburg PA
CBHW060825270326
41931CB00002B/67

9 781734 042511